Up

Up

Man's First Flight

Being a Tale of Genius and Courage, Luck and Skill,
Foolishness and Frenzy, Portent and Presumption

The Glorious and Goofy Genesis of Manned Flight

Darwin O'Ryan Curtis

Library of Congress Control Number: 2010901021
ISBN: Hardcover   978-1-4500-3226-1
                  Softcover   978-1-4500-3225-4
                  Ebook   978-1-4500-3227-8

This book was printed in the United States of America.

To order additional copies of this book, contact:
Xlibris Corporation
1-888-795-4274
www.Xlibris.com
Orders@Xlibris.com
73300

CONTENTS

Acknowledgements

ACKNOWLEDGEMENTS

Much of the background information in this book was provided by historian Marie-Hélène Reynaud, an authority on the Montgolfiers. The organizers in Annonay, France of the 200th anniversary celebration of the first manned balloon flight were also most hospitable and helpful. Our erudite friend Jacques Degas provided guidance. Ben Franklin's eye-witness accounts added that "you are there" touch. My wife Ann was with me gathering lore and got full enjoyment from the high-jinks reported here as she always did wherever high-jinks were to be enjoyed. The cover art is by Matthew Curtis; preparations for printing were made by Tracy Malloy-Curtis.

CHAPTER ONE
"Stock Up on Taffeta and Rope"

Some astronomers question that our universe began with a big bang. There is no room for doubt about one thing that did begin with a big bang: Man's venture into the air.

After the first balloon flight at Annonay, France in June 1783, balloons quickly became a fad. The fad exploded into a craze. Kings, commoners and peasants were all possessed. After a couple of years, and almost as suddenly as the fad began, it was over. People turned to other things, revolution for example.

A few of the first balloons were remarkably well conceived and built, but most were little more airworthy than lobster pots. Some of the people who built them and flew them were geniuses, some were fools. All, of course, were heroes when they succeeded and bums when they failed. Bums, heroes, geniuses, fools, they dealt with some of the most daunting problems that explorers have ever confronted.

"This invention was received with an enthusiasm never equalled in the history of inventions," said Dolfus and Bouche in their 1943 Histoire de l'Aeronautique. Well, maybe. Nobody measured popular reaction to the domestication of fire or the launching of the first boat or the debut of the wheel. It could be argued, though, that there hasn't been as much enthusiasm about a scientific breakthrough since manned flight, unless it was for the brief euphoria when Lindberg flew the Atlantic in 1927 or Armstrong took his "small step for mankind" on the moon in 1969.

Little wonder. The events of 1783 were astounding: man had freed himself from the timeless grasp of gravity. All subsequent exploits in air and space are but extensions of this phenomenal feat. And it is doubly astounding because two people did it simultaneously, independently of one another and by employing different means. Michel-Joseph Montgolfier harnessed the lifting power of hot air in a balloon; Jacques Alexandre Cesar Charles used hydrogen.

Joseph Montgolfier (1740-1810) and his brother Etienne (1745-1799) were the 12th and 15th children of Pierre Montgolfier, a paper manufacturer, and Anne Duret. Historian Marie-Hélène Reynaud writes that Joseph was a very imaginative but absent-minded party who ran away from school to the shores of the Mediterranean, there to subsist on shellfish. He did his own thing at the edge of indigence until he was 30, then married, set up his own paper factory and had five children. Even such sobering experiences were not enough to keep Joseph firmly in the real world. His vagaries got him into all sorts of financial problems and even a short hitch in jail for contempt of court, just a year before he and his brother flew their miraculous machine.

Etienne proved the perfect complement to his visionary brother. Tireless, cool and methodical, he studied science and architecture in Paris and made contacts there, including Ben Franklin, who were to be invaluable.

Joseph's first inspiration is said to have come while watching a shepherdess he encountered during his wanderings as she blew soap bubbles to catch the spirit of her departed mother. "If only bubbles," he thought, "could be solidified . . ." While visiting Avignon in November 1782, Joseph climbed out of bed into a chilly room one morning and had the peculiar idea to warm his shirt before putting it on. He lit some paper on the hearth and enclosed its flame in his shirttails. The story is, the hot air he captured caused the shirt to rise. (This experiment is best repeated with somebody else's shirt.) Joseph then fashioned a flared bag out of taffeta and when he held that over a flame, it climbed to the ceiling. "Can it not be said," asks Marie-Hélène, "that this was the moment of invention of the hot air balloon?"

Joseph wrote his brother, "Stock up on taffeta and rope, you're going to see the most amazing thing in the world." Well, the brothers fooled with Joseph's discovery for a month or so, unnecessarily complicating things by trying hydrogen and other gasses which escaped their bags. Then they burned wet straw and wool under a bag on the theory that the neutral vapor from the straw mixed with the alkaline gas from the wool would cause lift. This experiment worked, but for the wrong reason. It wasn't the gas that floated the bag; it was simply the hot air.

Before the end of 1782, the brothers succeeded with a couple of experimental flights of aerostats, but the big moment was to be six months later. On June 4, 1783, local government officials assembled in Annonay's Place des Cordeliers for the first public experiment. "Gentlemen," announced Joseph, "we will fill this large sack with a vapor of our making and you will see it rise to the clouds." Thirty-seven feet high and filled with the stinking heat of a wet straw and wool fire stoked by eight men, the aerostat rose. Afterwards, the boggled officials certified:

> The assembly having been invited yesterday afternoon to the testing of the aerostatic device discovered by the Montgolfier brothers . . . they found a vessel of about 28,000 cubic feet capacity built of cloth and doubled on the inside with several layers of paper, strengthened with a quantity of rope, a few pieces of wood and some wire. This globe, after having been invisibly charged, rose to the great astonishment of the spectators with increasing speed to the height of [3000 ft.] and having remained in the air about 10 minutes, descended slowly to earth at a distance of two miles from its point of departure.
>
> Since this discovery could prove useful, [the assembly decided to make official record of the experiment] which can only do honor to [the] inventors.

"An ancient dream of man had been realized," writes Marie-Hélène, "gravity had been conquered. An indescribable enthusiasm seized all. In Annonay, celebration followed celebration for several months[.]"

CHAPTER TWO
An Impeccable Witness

Fortunately for historians, Benjamin Franklin was in Paris at the time. He had arrived two years before with John Jay and John Adams to negotiate a peace with Great Britain after our Revolutionary War. Franklin stayed on as American Minister to the Court of France. He was 77 years old in 1783 and among his many accomplishments, was one of the world's most respected scientists. He followed the balloon experiments closely and reported on them to Sir Joseph Banks, president of the Royal Society in London. Once the experiments of Charles and Montgolfier became public knowledge, the competition between them sharpened and probably inspired both to greater effort. Franklin thought so. He wrote in November 1783:

> [T]he Emulation between the two Parties running high, the Improvement in the Construction and Management of the Balloons has already made rapid Progress, and one cannot say how far it may go. A few Months since, the Idea of Witches riding thro' the Air upon a Broomstick, and that of Philosophers upon a bag of smoke, would have appeared equally impossible and ridiculous.

Sometimes, the competition became acrimonious. When Etienne Montgolfier, Joseph's brother and close collaborator, turned up to honor Jacques Charles before his successful gas balloon launch of 27 August 1783, he got the bum's rush on Charles's orders. But neither of these serious scientists had much time for pettiness; they were preoccupied with physical problems never confronted before:

Can man survive in the air? Remember Ham, the smiling chimp who proved the feasibility of sub-orbital flight back in 1961? His experimental antecedents were the rooster, the duck and the sheep, which Etienne Montgolfier hooked onto the balloon he demonstrated for Louis XVI at Versailles on 19 September 1783. They came out of the ordeal all right except that the rooster got kicked around a lot by the sheep.

Can a balloon be steered or propelled by man? Etienne produced an abstruse mathematical formula proving the affirmative. Joseph convinced him with simple logic that the air resistance offered by the vast surface of a balloon could not be overcome by the force of man.

How about controlling balloon flight with sails? No good because the balloon itself acts as a sail. The smaller, conventional items would simply hang limp.

How can hot air be efficiently contained? Joseph saw this as a most pressing problem. The porosity of his first envelopes drastically limited the distance of flight. Small wonder, they were made of panels buttoned together. Jacques Charles had much more success with his balloon.

Franklin reported it was made of "silk being impregnated with a Solution of Gum elastic in Lintseed Oil, as is said. The Parts were sewed together while wet with the Gum, and some of it was afterwards passed over the Seams, to render it as tight as possible."

CHAPTER THREE
The Charlière

The balloon fever had started in Annonay, the Montgolfiers' home town south of Lyons, on 4 June 1783. This was the day of the first officially witnessed flight of a hot-air balloon called a *montgolfière* after its inventor. This was followed closely by Charles's first hydrogen gas balloon flight from the Champs de Mars in Paris on August 25th, just 82 days later. Gas balloons have since been known as *charlières*.

Charles had raised the money for his balloon by subscription at a bar. Its diameter of only 13 feet was a function of the clients' generosity. As it turned out, that was plenty big. Hydrogen had been identified by English physicist Henry Cavendish only 17 years before, and French chemist Antoine Lavoisier had just given it its name. It had only been made in tiny quantities by mixing vitriolic acid with water and iron filings. Charles needed a lot of the stuff, so he put iron filings in a wooden barrel about four feet high with two openings in the top. To one he fitted a pipe with a valve which conveyed the gas to the balloon. Water and acid were poured into the other opening. The trick was to dump in the liquids, then slap a plug into the hole to prevent the resulting gas from escaping.

There were problems. The valve controlling this highly volatile gas got so hot it had to be wrapped in rags to be touched. Escaping gas choked the crew. Water vapor rose into the balloon, condensed and ran out as a corrosive liquid which might have been man's first success at creating acid rain. After 13 hours of sloshing around in water and acid, the crew had only made enough gas to fill a third of the envelope. By mistake, a lot of *that* was lost. Well, it took 48 hours to fill that balloon.

When Charles's balloon finally rose above the walled courtyard of his house on Place des Victoires in central Paris, it had to be pulled back fast, to protect it from the public. Foot and mounted guards struggled all day to control a delirious mob of balloon fanatics. Charles was obliged to move the inflated balloon by dead of night to the launch site halfway across Paris. But first, he had to squeeze it out of his courtyard. He had, so to speak, built his boat in his basement.

Charles spent the day at the Champs de Mars topping up the balloon, fending off even bigger mobs and eyeing the menacing clouds. Finally, at five in the afternoon, two cannons were fired to signal the balloon's departure, and depart it did. Franklin was watching and reported:

> [T]he Globe was seen to rise and that as fast as a Body of 12 feet Diameter, with a force of only 39 Pounds, could be suppos'd to move the resisting Air out of its Way. There was some Wind, but not very strong. A little Rain had wet it, so that it shone, and made an agreeable Appearance. It diminished in Apparent Magnitude as it rose, till it enter'd the Clouds, when it seem'd to me scarce bigger than an Orange, and soon after became invisible, the Clouds concealing it.

For forty-five minutes Charles's balloon flew over Paris and out into the countryside before rising to an altitude which caused it to burst. It fell into a field by the little town of Gonesse situated near what is now the Charles de Gaulle airport. Two passing monks proclaimed it a monster and fearless peasants thereupon fell on it with pitchforks and scythes. The remains were tied to a horse and dragged triumphantly into town. There, the priest noted that the stench it exhaled through its wounds were characteristic of the devil.

Charles had been obliged to solicit funds for his balloon a copper at a time, surmount the perilous problem of filling it with hand-crafted hydrogen, defend it from successive mobs, launch it into a storm and then have it mindlessly shredded upon landing. Never mind, he was elated. With the Montgolfiers breathing down his neck, he turned immediately to preparing his first manned flight.

CHAPTER FOUR
The Montgolfière

Meanwhile, Louis XVI had gotten curious about the Montgolfier brother's Annonay experiment and invited them to demonstrate their machine at the Palais de Versailles. Joseph Montgolfier was too reclusive to get involved in that, but Etienne saw it as an important break because royal interest might lead to royal patronage. The date was set for 19 September. Etienne worked two months at his friend Reveillon's wallpaper factory in Paris to build a balloon worthy of the King. The envelope was made of paper glued to linen. Nothing was spared on its decoration. Etienne was ready in good time.

(The design of this hot air balloon was perilous to a fault. It was comprised of that very large and flammable balloon open at the bottom, with a platform-like gondola suspended underneath. Lift was provided by hot air from an open fire of straw built in the center of the gondola under the balloon's opening. The first trick was to get the air inside that envelope hot enough to fill the balloon and raise it above the flame before the whole contraption became an inferno.)

Since Montgolfier's balloon had been funded by the Royal Academy of Science, he felt it would be appropriate to give his benefactors a tethered preview. The academicians were invited for September 12th, an overcast, blustery day altogether unsuitable for exposure of so fragile a device with so exalted a destiny. However, Etienne didn't want to disappoint the august scientists or the thousands of balloon enthusiasts also on hand. He must also have felt the pressure of Charles's successful experiment three weeks before and feared the ridicule that a common sense decision to postpone might provoke. He decided to go.

Out of the hangar came the big, beautiful machine and his efficient crew of 20 had it filled and straining at the ropes in ten minutes. The members of the academy gaped, applauded and lined up to congratulate the triumphant Etienne. Then the squall hit. The crew's frantic struggle to save the balloon was in vain. The wind tore it and the water shrank it; the paper glued to it soaked off. Eight weeks of work was reduced to half a ton of trash in a moment, and the Versailles date a scant week away.

Etienne's numerous friends pitched in to put together another balloon in only five days. It wasn't doubled with wallpaper and it didn't have as fancy a decor, but there it was, finished, on the morning of the 18th. The trouble was, it was still raining, so the balloon couldn't be tested. (The French deplore the extremes of weather this time of year. They say September either destroys the bridges or dries up the fountains—1783 was a bridge buster.)

In the afternoon the weather seemed to clear a bit so Etienne got the balloon out for a test. When it had been hoisted into position above the flame, another squall hit, rending the envelope. It was dragged back in the hangar, hurriedly patched, and stuffed into a cart for the trip to Versailles. The 19th was yet another blustery day. The wind hampered filling and knocked some air out during the flight, but the balloon rose to about 1500 feet with its mini-barnyard and traveled two miles in eight minutes before landing gently, its envelope rent again.

On 21 November, only two months and two days after the Versailles demonstration, Etienne Montgolfier was ready to attempt the first free flight of man. The balloon he had constructed for the purpose was much bigger, 76 feet high and 46 feet in diameter. It could contain 2,200 cubic meters of air as opposed to 1,400 cubic meters for the Versailles model. It weighed 1,600 pounds.

The young and enthusiastic physicist Pilatre de Rozier was to be the pilot. He would be accompanied by a minor nobleman, Major the Marquis d'Arlandes, who saw in this adventure the chance for some major bling at court.

CHAPTER FIVE
Breaking the Gravity Barrier

The King, having figured the odds, decided not to attend this experiment. He did not want to be associated with a failure. On the other hand, if it did succeed, he would want the credit for having sponsored it. His solution was to send his son to represent him. Louis-Charles was two.

During most of the preparations at La Muette to the west of Paris, the weather had remained foul and caused endless trouble. Even at the moment of takeoff, a gust of wind ripped open the cloth envelope in several places causing it to collapse, narrowly escaping the flame of its heater. But on that cold November afternoon, women spectators kneeled on their coats and sewed up the rents. In an hour and a half, the balloon was ready again.

The balloon rose, the crowd roared to the waves of the first aeronauts. They were somehow to stay aloft for 50 minutes, rising and dropping vertiginously between 3,000 feet and the rooftops of Paris. They traveled about six miles, boxing the compass from SE to N to SW to S to E to SE again at the whim of the violent squalls. Their fire made several holes in the fragile envelope and burned through some of the ropes lashing the platform to the envelope. D'Arlandes put out a blaze that was eating at the gallery itself. A nobleman, d'Arlandes was probably disinclined to participate in the common labor of stoking the fire. Furthermore, he'd had enough. While de Rozier scrambled to keep his flying hibachi aloft, d'Arlandes was ardently advocating a hasty landing.

When land they did, near what is now Place d'Italie, only two and a half miles from their point of departure, d'Arlandes stepped out smartly to safety while de Rozier got pinned under the envelope as it collapsed into the flame. Managing to escape with his life, he was set upon by a mob, his coat torn off and ripped up for souvenirs. D'Arlandes hustled back to La Muette to steep in the acclaim of assembled dignitaries, while de Rozier, because he was now improperly attired, went home.

Franklin reported that: " . . . the Marquis d'Arlandes did me the honor to call upon me in the Evening after the Experiment, with Mr. Montgolfier the very ingenius Inventor. I was happy to see him Safe. [d'Arlandes] informed me that they lit gently without the least Shock, the balloon was very little damaged." De Rozier would have told a different story.

CHAPTER SIX
Going First Class

O nly ten days later, Jacques Charles and his confederates the Robert brothers were ready with *their* first manned flight. The site of the launching was the elegant Tuilleries Gardens in front of the Louvre. Charles and Noel Robert comprised the crew. Charles and the Roberts had devised a valve to release gas from the envelope to descend and took sand ballast they could jettison by the spoonful when they wished to rise.

A light breeze and a cloudless sky offered a rare perfect day for ballooning. Charles's troubles, for troubles he had, were elsewhere. Making hydrogen, for example. A lot more was needed than for the previous unmanned flight. Instead of one barrel in which to create their gas, the Robert brothers set up a circle of barrels around a pond. Tubes led from each barrel to a large bell which trapped the gas. The balloon was then filled through a pipe from the bell. Slowly.

They started filling on the 27th of November with the expectation they'd go on the 29th. Having badly miscalculated the time it would take, they had to announce a postponement until December 1st, to the disappointment of their contributors, an impatient, suspicious and generally unsympathetic lot. The king was big trouble, too. The morning of the 1st, with the balloon filled and ready, Charles got word that the king would forbid the flight. He considered it too hazardous. (Who could fault his judgment after the terrifying experience of de Rozier and d'Arlandes?) Frantically, Charles appealed, but the king was unavailable. The morning passed.

Rumors swept through the vast crowd packed into the center of Paris: the Montgolfiers were conspiring to frustrate Charles's experiment; the experiment was a fraud, etc. The mood of the subscribers, paying spectators and the encircling hoards turned from impatient to volatile to ugly. The pressure of this Roman circus on the scientists attempting such difficult and dangerous experiments was crushing.

Noon passed with no word from the king. Charles decided to go in spite of the king's injunction. He felt his commitment to the public transcended obedience to Louis. He might also have reckoned flying was the only way out of his predicament. He prepared to release a small pilot balloon to check the winds. Etienne Montgolfier appeared again to lend support to the experiment of his rival. Charles handed him the scissors to cut loose the little balloon. There was thunderous applause for this graceful gesture which erased Charles's previous insult to Etienne.

At 1:15 p.m. the aeronauts climbed into their gondola. It looked like a very small galleon with its high prow and poop. It rose "majestically," said Franklin, before that adjective became a cliché to describe all balloon risings. He estimated it leveled off at 2,000 feet. It was carried toward the north until it appeared "no bigger than a walnut, then disappeared." Charles, as Franklin pointed out, had the "Leisure to make Notes." Here they are in part:

> Nothing will ever equal the moment of sudden gaiety which carried me away when I felt myself leaving the earth; it was more than just pleasure, it was euphoria. To this sensation I soon added another: admiration for the majestic spectacle which was unfolding.

He added later:

> While I was making careful observations of the barometer, Monsieur Robert inventoried our riches. Our friends had ballasted our carriage as for a long journey: Champagne, blankets, furs, etc . . .

(It is a ballooning tradition to this day to carry a bottle of Champagne in the gondola. This is given to the farmer owning the property on which the balloon lands so he can properly celebrate the occasion in spite of the violence done to his crop.)

The aeronauts landed gently two hours and five minutes after takeoff, about 22 miles away (near where you will find now, of all things, the Complex Sportif Yuri Gagarin). Since the balloon was still airworthy, Charles took advantage of the remaining daylight to go up once more, alone. He rose to over 10,000 feet and made another perfect landing 35 minutes later. He saw the sun set twice that day, he said, once while in a vale where he had first landed and once, after rising into it again, when it disappeared below his extended horizon. He had noted the progressive fall in temperature as he rose, and the varying directions of the wind at different altitudes.

Charles and Robert had flown four times as far as de Rozier and d'Arlandes, gone over three times as high and stayed up over twice as long. They also enjoyed a certain serenity denied their rivals. However, it had taken them four days to fill their envelope, not to mention the expense, vs. ten minutes for the montgolfiére.

When Charles got back to Paris that night, he was met by the Marquis de Lafayette, recently returned from America, who did him the honor of driving him home.

Chapter Seven
The Wild Blue Yonder

In her book *Les Freres Montgolfier et Leurs Etonantes Machines*, historian Marie-Hélène Reynaud mentions that the intellectual salons of Paris were haunted by questions of whether Montgolfier or Charles had the greatest merit. "Franklin," she writes, "saw [the matter] very clearly behind his legendary little glasses: Montgolfier is the father of the balloon, Charles the wet-nurse." Anyway, both were elected to the Academy of Science. In fact, both got a lot of acclaim for their spectacular experiments.

The first big ballooning event of 1784 was organized by Jacques de Flesselles, the Administrator of Lyons. He saw in Joseph Montgolfier a means of achieving fame for himself. De Flesselles was good at organizing a subscription drive, if not at subscribing, and he lured Joseph to his purpose. (In the long run, Joseph had to pay for a lion's share of the experiment, cost overruns not being from yesterday.)

"The Flesselles," for the Administrator contrived despite his modest contribution to give the balloon his name, was gigantic. Ninety-four years would pass before there would be another as big. Made of two layers of light cloth lined with paper, it was 105 feet high, 93 feet in diameter and could contain 23,000 cubic meters of air. Joseph hired a ground crew of

150 to assemble and handle it. It took a week to put together after which it was too big to shelter from the elements.

Le Flesselles

Elements. Lyons could cop the booby prize as the site for balloon launchings in January. It has notoriously wet winters. However, it was on the 15th of January 1784 that the monstrous machine was filled for testing. It stayed up almost half an hour on its tethers that day, while lifting a ton-and-a-half of ballast. The sight of it provoked the Lyonnais to frenzy. That night it rained. The rain froze in a sudden cold wave severe enough "to split rocks," as the French say. The next day, the 16th, the balloon was stiff with ice. The bonfires Joseph's crew lit to melt it got out of hand and burned the crown of the envelope. Repairs of the fire damage weren't completed until the night of the 17th. That night it snowed; the 18th was spent shoveling out. On the 19th, the temperature was around 23° Fahrenheit and the envelope was again "hard as a plank." Joseph had to rally tailors and seamstresses to sew up all the new rents. By noon the envelope was filling.

Joseph was going on the flight with Pilatre de Rozier as pilot. Four noblemen had also been granted places which they had demanded as a birthright. With the balloon so much weakened by the punishing weather, de Rozier and Montgolfier sought to lighten its load by inviting the passengers off. Nothing doing. There ensued an absurd hassle during which the blue bloods threatened violence to defend their right, despite the knowledge they were superfluous and their weight would compromise the experiment. Incapable of prevailing on these fatuous Fauntleroys, de Flesselles ruled they might go, thus condemning any slight chance of scientific success. To frost the cake, a young salesman named Fontaine leaped into the gallery at the last second, adding his weight to the burden and his name to the record books as the first aerial stowaway.

The wounded, overladen balloon was finally loosed, but instead of "rising majestically," it dragged its gallery inelegantly off the launching platform and across a field as the band played. It caught on a fence post, got free and rose reluctantly to 90 feet from which height it headed straight for the nearby Rhone River. Joseph stoked the fire. The Flesselles climbed, turned and drifted back to the launch area at an altitude of 2,300 feet. After it had hovered for 15 minutes, there was a sudden, sickening noise. The repairs to the burned crown of the envelope had given way. Long rents released the buoyant air and the battered balloon fell at increasing velocity into a marshy field.

Miraculously, no one was hurt. The Lyonnais, having seen that behemoth of a machine ten stories high hanging over them, were deliriously happy, even though it hadn't gone anywhere. Their balloon had flown! They lost themselves in raucous revelry. Joseph Montgolfier and Pilatre de Rozier weren't so moved. They knew little serious purpose had been served—and the fiasco had brought Joseph to the brink of financial ruin.

While Montgolfier and Charles were the father and wet-nurse of ballooning, de Rozier and Charles the first pilots, the man who would become by far the most experienced pilot during those early years was another Frenchman by the name of Jean-Pierre Blanchard. By the time he was finished, he had chalked up more than 60 flights. Back in 1781, Blanchard had built a flying boat with eight sails, a gondola, oars and a rudder, but nothing had been invented that would lift it. Even so he kept his audiences enchanted demonstrating his contraption and reading discourses on flight. Understandably, Blanchard was considered a bit peculiar in those days. The great mathematician Laland ridiculed him in the *Journal de Paris* of 28 May 1782, pointing out it had been proved impossible for man to fly. That was just a year before the Montgolfiers put up their first balloon at Annonay.

Jean-Pierre Blanchard

By March 1784, Blanchard had gracefully recognized the genius of Montgolfier and added lift to his flying boat in the form of a charliére, an elaborate machine he had fitted with wings, rudder and a parachute to navigate himself through the air. As he was preparing to leave from the Champs de Mars where Charles had sent up his first balloon four months before, the usual crowd appeared. Suddenly, one Dupont de Chambon, a student at the nearby École Militaire, tried to clamber into the gondola, claiming the king had sent him. When restrained, he lay about with his sword, piercing the envelope, cutting Blanchard's hand and breaking some of the fancy furniture.

Blanchard's Balloon

Blanchard went up anyway for an hour and a quarter in his crippled machine and tried unsuccessfully to steer and propel it. He was to conclude, as Joseph Montgolfier already had, that man didn't have the strength to maneuver a balloon, and postulated a steam engine to do the job! (A year earlier, one Bulliard had suggested the retroaction of a rocket as a means of propulsion and made some successful experiments using a wagon. Thus, mechanical and even jet propulsion were proposed for propelling aircrafts at the very beginning of the age of flight.)

As the year 1784 unfolded, ballooning events proliferated. The beauteous Madame Tible became the first woman to fly. That was near Lyons on the 4th of June. Pilatre de Rozier and a colleague set records for altitude (11,000 feet), distance (31 miles), and speed (36mph) on the

24th. The Robert brothers flew a sausage-shaped balloon at Versailles on July 15th and didn't have any luck either trying to steer the thing.

Four days before that, one of the more ludicrous experiments took place in Paris. A publicity-seeking priest, the Abbé Miolan, built a balloon funded in part by his friend Madame Janinet, the wife of an engraver. Not just any balloon did he build, this was a balloon with differences. Already, new gimmicks were necessary to attract funding as well as the mobs of easily jaded Parisians.

Miolan's balloon was a big montgolfiére—a lot bigger than anything Parisians had ever seen, being over 100 feet high and 84 feet in diameter. Furthermore, he had done away with a launch platform which protected the balloon from the fire during filling. He substituted four long masts attached to the balloon's middle to hold it above the flames. He also added a vent in the envelope for propulsion, an enormous fan-shaped rudder and two little satellite balloons that were to fly above and below. These novelties served well to stimulate the sale of tickets.

Miolan would make his flight on a Sunday with his friend Madame Janinet (The aforementioned beauteous Madame Tible wanted to go too, as did many other ladies.) Marie Antoinette was scandalized when she heard that a man of God would organize such a carnival on a Sabbath, particularly since the dictates of decorum would prevent her attendance. However, she was mollified on hearing that the Chief of Police, in his pious wisdom, had picked the day of rest to prevent absenteeism.

With Swiss guards in place to restrain the restive crowds, Abbé Miolan began inflation of the enormous envelope which required the building of a regular conflagration. By the time it was half full, the Swiss guards had to be withdrawn to prevent their immolation. In spite of the inferno, the fatally heavy envelope sagged slowly onto the masts which impaled it, then settled quickly into the unshielded flames. The crowds, furious at the failure and no longer kept at bay, ripped up the debris and fed the flames. Their rage unsated, they threw onto the pyre the chairs set out for the paying audience. Then they looked for the good Abbé, but he'd finally done something smart: he'd vanished.

Down in Bordeaux, an unsuccessful launching so enraged a mob that two people were killed after which two were hanged and nine others sent off to row in the galleys. None of this was serving science. It was, rather, lighter-than-air madness that galvanized adventurers, exhibitionists, hustlers, machos and assorted crazies like a Frenchman named Bouché. He went to Spain to demonstrate a *montgolfière* on the first anniversary of the Annonay flight. He got off the ground, but a tether remained fastened to the crown of his balloon. It suddenly inverted, pitching him into the envelope with his fire. Somehow, he survived too.

Astonishingly, there were no fatalities at all during this silly season of ballooning. Ignorance was bliss. Balloonists proved impervious to the ravages of rain, wind, snow, ice, fire, angry mobs, superstitious peasants, presumptuous aristocrats, souvenir hunters and even their own dementia. Many instant celebrities like Miolan and Bouché returned quickly to the anonymity they deserved.

Chapter Eight
Mister Up-and-Down

On the other hand, Jean-Pierre Blanchard was a kind of Johnny Appleseed of ballooning, traveling widely to sow interest in the miracle of flight. He is credited, for example with the first flights in Germany, Holland, Belgium, Switzerland, Poland and what is now the Czech Republic. Later, he turned up in Philadelphia carrying a supply of sulphuric acid with which to brew his own gas. On 9 January 1793 he made the first ascent in the New World in the presence of the president of the United States. George Washington provided him with a personally signed passport requesting assistance to the bearer wherever he might drop in.

Blanchard's most famous adventure, and it might well have been his hairiest, was crossing the English Channel. This was the first flight across water and also the first flight for an American. Blanchard had agreed to take Bostonian Dr. John Jeffries with him, on condition Jeffries pay the costs. They left Dover at 1:05 p.m. January 7, 1785, against the advice of local sailors and to the great distress of Mrs. Jeffries. The balloon was a leaky *charlière*. The basket was laden with life jackets, food, clothing, a supply of pamphlets about ballooning, scientific instruments, a propulsion system and rudder (despite his previous conclusion they wouldn't work), two anchors, ornaments and ballast. Nevertheless, it gained altitude for the first 50 minutes of flight. After that, it began an inexorable descent.

The ballast went out first in an attempt to maintain altitude, then the pamphlets. Half an hour later they were followed by the food, the propulsion system, the rudder and even the ornaments. As the descent went on unchecked, both anchors were jettisoned. With further loss of altitude, Blanchard and Jeffries threw out their clothes and put on the life jackets.

The naked aeronauts in their empty basket were now three-quarters of the way across, watching with mounting dismay the relentless approach of the waves. Suddenly, Blanchard got an inspiration. It had that luminous logic of which the French are so justifiably proud: Jeffries must jump.

The issue was never debated because at that moment the balloon began to rise. At 3 p.m. the French coast was crossed and the balloon, now somehow too ascendant, had to be bled of gas. Blanchard and Jeffries finally wrestled it down by grabbing at passing trees in the forest of Guimes where a column was later erected to commemorate their feat. It is still there. You may also see the basket at the museum in Calais. It remains empty.

CHAPTER NINE
"The First Man Who Left the Earth"

W hile Blanchard was in England in 1784 preparing for his channel crossing, Pilatre de Rozier was busy becoming a favorite at the court of Louis XVI. As Reynaud puts it, "he was no longer considered by the great personages as the valet of Major d'Arlandes." From Monsieur de Calonne, the comptroller of the mint, de Rozier obtained a grant for "the glorious mission of crossing the Channel by air."

Etienne Montgolfier had already designed and built a balloon for this purpose, but de Rozier had appropriated it, cunningly named it the *Marie Antoinette*, and flown it out of Versailles as Louis' entertainment for the King of Sweden. It had burned on landing, as *montgolfières* were wont to do, and peasants had savaged what was left, as peasants were wont to do. So de Rozier had to start over. In building the new machine, he decided to marry the *montgolfière* and the *charlière*, profiting from the advantages of each. He made a gas balloon with a lift equivalent to the weight of his craft. It was, of course, only a fraction of the size of a montgolfiére because hydrogen, although highly flammable, has 14 times the lift of hot air. Under that, he affixed a cylindrical chamber in which he could heat air by a fire at the bottom. He reasoned he could control altitude more efficiently with a *montgolfière*-type fire than by the alternative releasing of gas and ballast *charlières* required.

Pilatre de Rozier

The hydrogen balloon was made of taffeta impregnated with linseed oil and doubled with a layer of cow gut stuck on with a secret formula of glue, honey and oil. This recipe rendered his envelope so tight it remained inflated for two months without showing a wrinkle. The hot-air cylinder was made of cloth with a chamois top. The inventors of the two balloons he combined were horrified. The Montgolfiers wrote him anguished letters pleading that he not risk his life in so dangerous a device. Charles told him it was like putting a spirit lamp under a barrel of gunpowder. Pilatre de Rozier, unmoved, went off to Boulogne with his the *aeromontgolfière*, as he called it, in late 1784 to await a favorable wind. There he was dumbfounded by the news of Blanchard's preparations for a flight from Dover to France. Given the persistence of westerly prevailing winds, de Rozier had the long odds.

Soon, de Rosier had to go to Calais and felicitate his triumphant rival. The channel having been crossed, de Rozier suggested to Monsieur de Calonne that their project be abandoned. De Calonne agreed, but on condition that de Rozier pays all the costs exceeding his grant. De Rozier didn't have that kind of money, so he stayed with it. He had another incentive, too. He had met a beautiful and wealthy English girl and hoped that by his success with his venture he might win her hand.

Five more months passed waiting for a favorable wind. Meanwhile, rats, hoards of rats, found the *aeromontgolfière* to be gourmet dining and tied into it. De Rozier conscripted a defending army of dogs and cats to little avail. Finally, he hired men to beat drums all night long to keep the rapacious rodents at bay. On top of that, he would have lost his machine to a howling storm, had not local officials come to its rescue.

All the while de Calonne, either through ignorance or malice or both, kept the pressure on de Rozier to go, the winds be damned. "The government has not," he wrote, "spent 150,000 francs for a physicist to voyage around the coast of Picardy. You must use the machine to cross the channel." On 15 June 1785, the wind finally seemed favorable. Pilatre de Rozier and his collaborator Pierre-Ange Romain got away after having to apply the usual stiff-arm to an arrogant marquis asserting his right to ride in other people's balloons.

The *aeromontgolfière* went out over the Channel while rising to 1700 feet, veered and returned to the French coast. After 27 minutes of flight, the predictions of the Montgolfiers and Charles came to pass. The machine was seen to catch fire and it fell at Wimereux. In nearby Wimille on the Paris-Calais road, there stands a tomb inscribed to "The First Man Who Left The Earth."

CHAPTER TEN
Ballooning's Godfather

It was around this time that public delirium over balloons cooled. Now it was the turn of serious people to perfect the inventions of Montgolfier and Charles and find practical applications for them. Franklin encouraged this effort and even offered some ideas of his own such as:

> [E]levating an Engineer to take a View of an Enemy's Army, Works, etc., conveying Intelligence into, or out of a besieged Town, giving signals to distant Places, or the like.

And further to military applications:

> Where is the prince who can afford so to cover his country with troops for defence as that ten thousand men descending from the clouds might not in many places do an infinite deal of mischief before a force could be brought together to repel them?

It will come as no surprise that all these ideas were realized, but that is another story.

Franklin had been an influential and eloquent balloon enthusiast since the very beginning. If Montgolfier was the father of ballooning and

Charles the wet-nurse, Franklin may merit the title of Godfather. Listen to him scolding Sir Joseph Banks about what he perceived as Britain's "Bashfulness" for fear of failure in experimenting with flight. He wrote it on the day Pilatre de Rozier first flew:

> This Experience is by no means a trifling one. It may be attended with important Consequences that no one can foresee. We should not suffer Pride to prevent our progress in Science. Beings of a Rank and Nature far superior to ours have not distained to amuse themselves with making and launching Balloons, otherwise, we should never have enjoyed the Light of these glorious objects that rule our Day & Night, nor have the Pleasure of riding round the Sun ourselves upon the Balloon we now inhabit.

www.ingramcontent.com/pod-product-compliance
Lightning Source LLC
Chambersburg PA
CBHW050348290526
45785CB00006B/2678